ESKI, ALEXANDER and PANDORA

INNOVATIONS of OSM

Dean LeRoy Sinclair (BA, MS, PhD)

Copyright 2018, Dean L. sinclair

ESKI'S GRAIL, ALEX'S SWORD, pANDORA'S BOX:

Innovations of OSM

Introductory comment: When this was started there was certainly no intention to run the list ubtil it became a small book. It started as an exploration of "How many little changes would the theory make? Just what kind of a Pandora's box was accidentally opened here by doing an "Einstein on Einstein?" The list of insights and innovations kept running into one another and led to listing of ideas that had not been published before.

This is not really a book for a beginner. My other books try to be that, It is a book for some thoughtful person who might be in a position of "real existence-in-the-socio economic-society of-scierne." might find a use for the information. There is literally a "Pandora's Box of ideas that -- if picked up on by someone in a position to be paid attention to--could cause change.

Much of this has been around im writings on the Internet for about a decade and in books published on CreateSpace since 2015 and in *"The Grail of Physics"* which my friends at Earth/nitruX had printed at Harvard Book Store. Charles william Johnson liked the book *"oscillators in a Substance Model of Existence"* so much that he offered to publish an Earth/matriX version at his own expense. Of course, I took him up on it.

This book is not intended to be a "quick-and-easy-read." It is double-spaced so that the reader will have space to mark the book up adding comments and questions, refutations and , maybe, kudus.

o

INNOVATIONS RELATED TO THE OSCIIIATORS IN A SUBSTANCE MODEL OF EXISTENCE

1. EINSTEIN'S RELATIVITY WORK MOVED FROM PHYSICS THEORY TO MATHEMATICS OF INFORMATION. RECOGNITION OF SPEED OF LIGHT AS AN AVERAGE VELOCITY OF INFORMATION CARRIERS (AND HENCE A MAXIMUM VELOCITY FOR INFORMATION OVER ANY SUBSTANTIAL DISTANCE.)
2. THE BEGINNING OF A FIELD THAT ONE MIGHT CALL "STUDY OF PERCEPTUAL UNIVERSES BOUNDED BY A MAXIMUM VELOCITY OF INFORMATION TRANSFER."
3. REVISION OF INTERPRETATION OF MICHELSON-MORLEY EXPERIMENT OF 1890 SO THAT SPEED OF LIGHT IS SEEN AS AN AVERAGE TANGENTIAL VELOCITY OF THE "AETHER" RATHER THAN A NEGATION OF THAT BASIC SUBSTANCE.
4. DEFINITION OF THAT AETHER AS A MATRIX OF SPINNING "DOTS" HAVING AN AVERAGE

TANGENTIAL VELOCITY OF THE SPEED OF lIGHT, "C."

5. FURTHER DEFINITION OF THAT MATRIX AS HAVING THE CHARACTERISTICS OF A SUBSTANCE AT ITS TRIPLE POINT.

6. DEFINITION OF LIGHT AS PRESSURE DISTURBANCES IN THIS SUBSTANCE.

7. DEFINITION OF A ONE BASIC FORCE AS THE AVERAGE PRESSURE WITHIN THE BASIC SUBSTANCE.

. 8. DEFINITION OF MASS AS THE PRESSURE WITHIN THE SURFACE OF AN ENTITY WITHIN THE SUBSTANCE, ALLOWING THAT ENTITY TO HAVE EXISTENCE <u>(IT MAY BE NOTED THAT THIS IDEA COULD BE TURNED AROUND TO REDEFINE BASIC FORCE AS BEING THE AVERAGE OF ALL MASSES AS SUMMED AT ANY GIVEN POINT AND TIME.)</u>

9. DEFINITION OF ENERGY AS " A PACKAGE OF MOTION." AS USUALLY USED, MEANING kINETIC ENERGY, MEASURED BY COLLISION AND THOUGHT OF IN TERMS OF POSSIBLE UTILITY TO HUMAN PURPOSES.

10. NOTICE MADE THAT THE INTEGRATION OF MOMENTUM TO ENERGY COULD ASO BE INTEGRATED AT CONSTANT VELOCITY TO FURNISH ANOTHER ENERGY EQUATION, $E=(VM^2)/2$; OR , ONE COULD INTEGRATE "P" TO obtain $(P^2)/2$; and, reinserting mv for p, obtain a "TRUE ENERGY EQUATION" OF $E=(M^2V^2)/2$,

11. DERIVE FROM THE ABOVE BY DIFFERENTIATING THE EQUATION WITH RESPECT TO MASS WITH VELOCITY CONSTANT AT "C" SO THAT IS OBTAINED AN OLD FAMILIAR FRIEND. MC^2, IN A DIFFERENT GUISE,

as a sort of momentum . as the change of mass at the velocity "C" Considering that energy is considered as one half momentum times velocity; then, perhaps, energy content at velocity, c ,would be, $mc^3/2$ and the collisions in the Hadron Collider of "accelerated protons" would actually be of entities having the energy of some 4.5×10^{30} amu. (This sounds more like a small "neutron star" than an "accelerated proton.")

12. As Planck's Constant and the Speed of Light are both related to electromagnetic radiation, their ratio, h/c, is a constant related to electromagnetism. or in the case here, would be related to the motion of the information carrying dots. This constant has the dimensions of work or energy per cycle.and also of torque (mass times radius.) Hence, this constant can be defined by the equation, $m \times r = h/c = r \times m$.

This equation could define a family of oscillators, Assuming "Rest Mass" to be involved in one limit, this idea was checked with the rest-nass data for the electron and the proton, in each case the "r" value was found to be the literature value known as the Compton Wavelength. The equation clarifies the particle /wave-duality problem. The left and right sides of the equation define particle entities while the constant clearly defines the energy of a vibrating entity.

13. As h/c has the dimensions of "Work per cycle," independent of the length of the cycle, it defines the basic Work Unit of Our Existence. it could be designated as "q," the basic Quantum which gives the name to Quantum

Mechanics and the Quantum Revolution of the Twentieth Century.

14. *It can be seen that the phenomena of anti-particles and charge should be closely related. As each particle is a mirror of the other they presumably differ in predominant spin orientation--clockwise or counterclockwise-- this predominant spin-orientation gives rise to the phenomenon of "Charge."*

14. a. *The antimers appear to be the "separated halves" of a parent oscillator having counter-rotating halves. .In the case of the electron/positron set there is known to form, what could be called a "proton-less Hydrogen Molecule." This unit, known as Positronium, exists for long enough to have some chemistry and identified ortho and para forms. Eventually this unit reaches orientations which can collapse to what we could call the "Chargeless-Hydrogen-atom, the Zerotron, or the basic e^0 unit."*

14. b. *It may be speculated--without any secure basis-- that these mirror-image pairs may be interconvertible or considered as phase-shifted forms of the same unit. One* *my even guess that the "electron-pair' of electron-pair-bonding is two of these units, or even the Positronium unit, prevented from collapse by interactions with the other units*

15. As these are mirror-image halves, they can recombine to form the "original" oscillator with loss of kinetic energy equal to mc^2. As this resulting parent oscillator is "spin neutral" --hence uncharged--it is essentially undetectable as a unit (except by certain effects that have not been ascribed to it.) The units are said to "annihilate,"   This gives rise to the idea that "Matter and Antimatter Annihilat on Contact." That the reverse process, the splitting of the base oscillator to the pair (a process called pair-production) is known, has never been considered evidence for the existence of the parent unit.

One might call the following a "Sidebar."

Other than for the Feynman-diagram "virtual electron," which is the same unit, mathematically; this e^0 unit, possibly the progenitor of all other masses, was not suspected before the Internet publication by this writer in which he dubbed it the

"Zerotron" as the intermediate, parent, combined form, starting and intermediate point for the Negatron and Positron.]

16, The splitting by sufficiently energetic radiation of this combined form, the Zerotron. accounts for the phenomenon of "pair-production."

17. The Zerotron may also be considered a probable major constituent of dark matter.

18. The definition of Mass as a force shows that the Center of Mass is not just a mathematical convenience, but an actual point of combined force, equal to the combined mass of the entire unit. This is a constantly moving "point force."

18. a. This constantly- moving-point will be surrounded by other summations, giving the illusion of there being a true mass in that center.

16, b. This constantly moving center acts as a "magnetic stirrer" having the effect of heating any closely coherent object, adding to its "core heat" and causing liquefaction of cores that would otherwise be solid. It accounts for the hot internal cores of such units as Pluto.

16. c. Moving Centers of Mass account for the volumes considered as Black Holes by astronomers.

16. d. As this moving "CoM" would also occur in accretions of "Dark Matter." it would account for the flattening of the centers, leading them to appear "cored" rather than "cuspy"[A somewhat flattened disc rather than with defined central peak.}

17, Going back to the equation, m x r = h/c = r x m, suggests some other things. One is that, if rest mass and Compton Wavelength are the dimensions for the smallest mass and the longest wave length, then it may be possible that switching the coefficients could define the largest mass and the corresponding smallest radius (wave length.) If so, the electron would vibrate through a distance some 1832 times as wide as that of the proton and could vibrate from far outside a proton (or nucleus) to far inside of either. The massive "- Partner" of basic units, which is postulated by certain theories would exist within the unit rather than "somewhere elsewhere."

18. Additionally, the above equation suggests that the square root of "q" (h/c) could be considered as an average unit--a value which the basic units would vibrate through, This would be an entity which could be considered as either "always there," or "never there," depending on the view point. As h/c ("q") has a value of about 2.21×10^{-37} gram-centimeters, this "SinFree Entity" would have dimensions of about 4.7×10^{-19} grams and centimeters.

19. These values could be used to define a postulated basic oscillator operating at the wavelength of 4.7×10^{-19} cm.

20. An interesting mythos arises as an alternative to the "'Big Bang" if one assumes that an oscillator defined as above operated to convert bits of the "Substance" into Zerotrons; then, later, causing shock wave impulses which convert the Zerotrons to neutrons....

21. The idea of the square root of a constant being another constant gives a thought about the square root of the Speed of Light. This value, about 173 Kilocycles/sec. is in the radio range, in an area noted, years ago, for

being "quite useless because of the amount of static." The exact value, however, if the guess be correct, could be a rather constant background frequency.

Additionally, another strange idea arises. Frequencies originating from transmitters of higher frequency would show a "redshift." as is well mown. However, transmitters operating at a lower frequency would show "*blue shift !*"

Insofar as the writer knows, these guesses have never been checked on, and seem as valid as ideas derived from complicated analyses of Einstein's work.

22. As protons and electrons would be subjected to the 'One Force" as well as the modifications of that force due to their motions, the nuclear atom model arises here, as an interior region defined by the clustering of the protons, the "nucleus," surrounded by an outer--and, perhaps, also, inner--sphere, defined by the motions of the electrons.

23. As the motions of the basic units, would, in general, occupy spheric spaces, it is possible that the standard

descriptions of electron orbitals actually describe packing patterns of spheres.

24. *Electromagnetic emission/absorption may be better understood as rearrangement within the entire atom rather than the movement of one electron one "orbit" to another.*

25. *The fact that motions can be considered as transferable among 14 directions--the six Cartesian Directions interacting with eight "cubic-corner" directions--may account for a number of observations, and suggests interesting pulsation model for a "Control Oscillator" which could have arisen from combinations of the e^0 unit.*

26. *Seeing the Zerotron as a possible "Hollow-Sphere" oscillator which could actually encase protons or the electron/positron units, suggests the possibility that the Zertron, e^0, unit may be not only the basis of "normal matter" but also the basis of "Dark Matter," and neutrinos.*

27. *Considering a One Force acting on rotating-inverting units leads to consideration that the electron-pair bond of*

,molecular chemistry may do as much to hold units apart as holding them together.

28. "If there be only one electron between two units, they may collapse to one unit" becomes an operating premise. For example, we might expect HH+ to collapse to D+, DH+ to collapse to T+, DD+ to collapse ro He+, OO+ to S+, etc. That is, diatomic cations be collapsed to their monatomic isomers.

29. The collapse of monocationic ions of Hydrogen molecules--taking place within the interior of metallic cations- *-gives a rather simple basic explanation for the phenomena known as "Cold Fusion."*

30. Gravitational-type effects of the One Force would be present within atoms. (A deuteron formed within a Nickel cation, would drop toward the nucleus for the same reason an asteroid falls to Earth.

31. Observation was made that all naturally occurring isotopes of elements could be considered as made up of Deuterium, Tritium and Helium Three.

This led to a useful view of nuclei being composed of units 3++, 3+, 2+, 1+, called "Four-factor coding." Using this coding and writing their numbers in the above order, natural Gold, Au (79. 197). codes as 13, 52, 1, 0.

32. It can be seen that any two units of the same atomic weight could be interconvertible through a "cation-in-common.' This would have its basis in interconvertibility of Tritium and Helium Three by moving an electron from "the outer pack" to an interior grouping. That is, the cation notated above as 3+, could be considered as the mono-cation of either Hydrogen 3, or Helium 3, This allows the ion to have some degree of flexibility as to the location of one electron, and, the positioning of the protons in the nucleus.

From their magnetic moments, it appears that Tritium is a flipping unit, while the Helium Three nucleus stays in triangular configuration.

33. As has been seen previously, this model indicates that basic units can combine particle and antiparticle to form a neutral unit, and it is possible that the two units are interconvertible. The theoretical form of each unit suggests that each spends some part of its cycle--a guess could be about 40%--as the opposite form. Compex units would not undergo the combination to a vanishing form which is possible for the electron/positron and "proton/'conton'" forms; therefore, the idea that "Matter and antimatter annihilate on contact," is an error.

33. a. As matter and antimatter do not annihilate, very simple units may interconvert.

33. b. It may be that matter and antimatter units cooperate such that what we call "matter" contains both antimatter and matter units.

34. A follow up idea on above. What we call "neutrons" in nuclei could actually be considered as "cooperating anti-matter," and charted in the same way that the matter component is charted; that is by the "electron-orbital charting," (1s2, 2s2,2p6,etc.)

34 .a.This charting, which often gives useful insights could also be considered as charting an inner set of electrons, not normally involved in chemical reactions. This is ptonsnly a more easily accepted alternative to the "cooperating-antimatter" idea.

34. b. Still another alternative which gives the same chart is to consider there be "imbedded electrons" in the

units chaired by the Four-factor Model where there is considered one imbedded electron in 3++ and 2+ units and two imbedded in 3+ units. <u>(This last view allows the Four -factor and the "Electron-orbital, dual-charting" ideas to be used together without any "intellectual conflict.")</u>

34. c. It can be seen that there are at least three alternatives to considering neutrons as being present in atomic nuclei.

35. Considering that the idea of neutrons in nuclei was a "jumped on conclusion " because neutrons were shown to exist, we need to see where they could come from in nuclear fission. The OSM model suggests two possibilities. One is that some of the ever-present, dark-matter Zerotrons are shock-wave transformed to neutrons. Another is that, if the Four-factor Model be correct, the commonest unit in most atoms--the 3+ unit-- may be easily shock-wave convertible to 2+ and n^0.

LET US START A NEW TOPIC: THE RELATION OF THE OSM THINKING TO THREE BETTER-KNOWN MODELS.

1. Time-Space Modeling/Minkowski-Einstein Space.

Space is usually measured in linear units such as the foot. Time is measured by units related to some reproducible cycle; in other words, in terms of a sequence, Time/Space is a strange hybrid; which, however, seems to work quite well in describing happenings within a substance made up of rotating dots-- a substance named "SpaceTime."

This approach could be called a "Substance Model, Without Oscillators."

2- Quantum Mechanics describes wave motions as quantized. It does not actually recognize the basic quantum which OSM identifies. It is a modeL closely related to the oscillator portion of OSM. but recognizes neither oscillarion nor a basic substance..

It appears that both Space -Time and QM would be "OSM compatible." They could be considered as limited, partial models of the more comprehensive Oscillators in a Substance Model.

3. The Standard Model of Particle Physics and the Oscillators in a Substance Model have one point of "sort-of-agreement' in the idea that there seems to be a hierarchy of levels of "basic" units. There is disagreement as to what falls into which levels and definite disagreement as to what are "basic" particles. The OSM attitude appears to be that there be one basic unit of the "Substance," Beyond this is likely one more, probably the unit called the "virtual electron" by Feynman Models, and the Zerotron in this writing. Derived from this, are the electron and proton--by way of the neutron.

OSM would place the negatron/positron , Positronium, and Zerotron in a first level, with the proton/conton unit and a set of related units in the second level and more

or less agree to having Muons as third level and the Tauons as the fourth.

OSM would list neurons as in a combination level and consider similar intermediate units to be possible.

OSM is compatible with unstable units--somewhat corresponding to molecular forms-- from level combinations.

However, the idea that the release of basic units by "atom-smashing," which is the basis of the Standard Model, simply does not fit into OSM reasoning. Neither does the concept of entities without mass. The idea of a particle which "bestows Gravity" is inconsistent with the simple definition of Gravity that arises from the One Force Concept found in OSM.

In this writer's opinion, if one works with the OSM concepts, most of the Standard Model of Particle Physics may be ignored, despite its impressive list of "Nobel-Prize-Certified" accomplishments.

SOME "FALL-OUT' ONTO MATHEMATICS.

1. <u>T</u>o generalize the basic-oscillator-defining equation , {m x r = h/c) so as not to be confined to one specific portion of space, it was realized that work should be done in absolute values rather than signed number values,

2. It was also noticed that the convention of assuming that a number without signs be signed positive numbers has the effect of placing models in the upper-right-forward octant of a set of Cartesian Coordinates, which could lead to errors as this ignores $7/8$ of 3-D space.

3. The conclusion is that this Convention of Convenience could cause some mathematical models to erroneous conclusions because working with only positive number results ignores negative number results, and "mixed number fields,"

4. <u>The above led to the realization that "Signed Numbers" should probably be considered as a subclass of vectors as they have both a directional sense specified by the "operators" (+ and -) and a</u>

<u>magnitude, designated by the absolute number to which the operator is attached,</u>

5. *The operators have different meanings in addition and multiplication. In addition function they represent a reversal in direction or sense of the operation. In multiplication they represent "right-angle' changes according to an automatic sense, with the changes being to-the-right, up and forward for positive change and to-the-left, down and bact for negative change.. These are in respect to Cartesian Coordinates which are related to human body preferences, and to human social conventions. Factors mathematicians use automatically.*

6. *Using signed numbers, Addition may be considered as a Multiplication system, operating in the second dimension of space, a line. Reversal be the only possible action, with operations confined to a line.*

7. *The above leads to a new type of definition for Zero in dealing with models of reality. Zero becomes the undefined-as-to-size-and-shape dot which is the*

starting point for any operation. It is the simple for the first dimension, the dimension of Existence.

8. The size and shape of the "Zero" for every subsequent action of a multiplication is therefore defined by the previous aeration. A "dot' is the defined unit of a line, a "straight" line-segment in the defined unit for a square, a square is the "Zero" for a cube, the cube is the Zero for a Line of Cubes, and so on.

8, It can be seen from the above that a signed number, generated on a set of Cartesian Coordinates, does not have a root, it has a history of diverse operations which define its position in space. However, this is not exact. For instance, were we to ask for the square root of + 4, do we mean that we want the number that would define the side of the square to the right and above the origin, or the one to the left and below? Similarly for the square root of -4 we would want to define one of the sides of the squares to the left and above or to the right and below. As these are "mixed-value" areas, it is claimed that there is no *"real" number that matches, so there is assumed an "imaginary number ", which is--unconsciously and automatically--considered "Positive;" which. multiplied by itself. will create a negative area.*

9. It may be contended that since they are necessarily 'numbers" which are beyond the scope of the two operators , + fand - . The 'roots" of signed numbers are all of a different form than are the signed numbers themselves, and since they are. apparently, as Is "i", considered to be a new unitary, posItive, system, then the roots of "real-signed-numbers" (signed numbers of the "positive field,") should also be considered as "imaginary." A conclusion may be reached that the field of Math called "Complex Numbers" needs complete revision of its basic terminology and revised to contain two sets of 'imaginary numbers," and reciting of places where absolute values are called"real numbers." Perhaps "r" could be considered as the "absolute value of the "square root of +1 as "i" is apparently considered the absolute value of the square root of -1.

10. It appears that there be a field of mathematics extending ideas such as noted above to other uses with what might be called "Ultra-complex numbers" dealing with the "ultra-imaginary" higher roots of signed-number-spaces and volumes and higher-order aggregates.

11. Defining Zero as the first Dimension, raises ideas about the number One, and the concept of Infinity. It is useful to remember that the number "One" is the symbol for "Wholeness"and can be of considerable utility in calculations from that standpoint. Similarly, one should remember that Infinity is the next number beyond where ever one stops counting or measuring, for whatever reason. It is not some incalculably huge number which suddenly appears, as it seems often taken to be. There are limitless numbers of stops in counting or measuring, an unlimited number of "Infinities." (The fun games of mathematical "Infinity" seem to be based on the fact--never mentioned, perhaps unrealized--that by Infinity being the next number, one unit beyond, rhere is being

admitted the existence of all of the previous numbers. no matter how they were counted.)

ENOUGH FOR NOW!

An Afterword: OSM and Cosmology

The Zerotron offers a possible explanation for Dark Matter. The Control Oscillator may be considered as an explanation for Dark Energy. The Control Oscillator may also prove to be the source of the Microwave Background which is said to be "left over from the Big Bang," (That idea, of course. is nonsense, if the "Big Bang" be a single event.) Some respected cosmologists, however, treat a line of events following a Big Bang episode as if they have been proven. What the proof is, I do not know. If the sequencing be correct then it would seem that the Big Bang would be a continuing event or series of events in order to account for the Microwave Background.

Possibly some genius Cosmologist/Mathematician could do a synthesis of the two ideas. It appears to this observer that such a synthesis might include the ideas of multiple "Bubble Universes", such as ours, and even

Bfanes and other dimensions.

:{The following is <u>not</u> republished from the book ,

:"Off the Wall, vol.2, Poemz")

A Thought on Mathematical Models

The mathematician/cosmologists among us
I suspect may agree:
"There is much more reality
than is to be found in the "Cartesian Octant
of 'Positive x, y and z.' "

Have fun.. ds

FIN

Finally?

PS. A list of my books published on Createspace may be

easily found by entering Dean LeRoy Sinclair in Google

Search.

Contact points, as of Jan. 2, 2018 :

deanlsinclair@gmail.com   and 1-605-290-2154

d

www.ingramcontent.com/pod-product-compliance
Lightning Source LLC
Chambersburg PA
CBHW060008230526
45472CB00008B/2006